A Boy Named Pluto

ASHLEY GALLEGOS AND
JAHN STOPPERAN

Copyright © 2016 *Ashley Gallegos and Jahn Stopperan*
Edition 1A

All rights reserved.

ISBN-10: 1539343227

ISBN-13: 978-1539343226

A BOY NAMED PLUTO

APPRECIATION

The authors would like to thank a variety of contributors who helped bring this book to life.

Jan Cray, Tom Stopperan, Jacqueline Lundeen, Tom Stopperan, Rian Stopperan, Debby Becker, Terry Haley, Tony Stopperan, Sue Stopperan, Teri Peppe, Dale Stopperan, Chris Garborg, and Colton Stopperan.

A BOY NAMED PLUTO

Table of Contents

Meet the Characters	2
Introduction	5
Chapter 1 The Bad News	7
Chapter 2 Getting Help	13
Chapter 3 Heading to Vinny's	19
Chapter 4 Marti Joins In	25
Chapter 5 Earl's Business Partners	29
Chapter 6 Vinny's Plan	33
Chapter 7 The Courtroom	39
Chapter 8 Support for Pluto	43
Chapter 9 The Arguments	47
Chapter 10 All Is Good	55
Did You Catch the Connection?	59

MEET THE CHARACTERS

Pluto is a kind-hearted child who dreams of an exciting future. He loves to spend time with his dog, Nix, and he is eager to grow up and finally fit in with the other planets in town.

PLUTO

Tooner is an outgoing and wild youngster who loves to play with Pluto. He can always be seen with flies circling around him, representing Neptune's 14 orbiting moons.

NEPTUNE

Ursula is a nurturing, soft-spoken woman who is excited to become a grandmother. Despite Uranus being the coldest and windiest planet, Ursula does her best to portray warmth and calmness to others.

URANUS

Ringer is a bipolar teenager who never leaves the house without his smartphone. The mood rings on each of his fingers make it hard for Ringer to hide how he is feeling.

SATURN

Jesse is the neighborhood jock and loves the Universe Wrestling Foundation. He is the largest figure in the group and he has a tendancy to intimidate people with his hot-air comments.

JUPITER

A BOY NAMED PLUTO

Marti is an energetic young woman who loves the adventurous outdoors. Her cold feet and white tuft of hair represent the iced poles of the red planet.

MARS

Earl is a mysterious old man with continuously changing cloud-like hair. While Earl may be friendly to his neighbors, he is dedicated to his career and likes to keep to himself.

EARTH

Vinny is an intelligent and well-liked computer geek who is best friends with Merk. He enjoys spending time indoors and is not afraid to take on a challenge.

VENUS

Merk loves to vacation near a beach and enjoys soaking up the sun. His go-with-the-flow attitude enables him to get along with everyone.

MERCURY

Eris is a little girl from Kupier, a region near Pluto's house filled with little girls and boys similar to Pluto.

ERIS

A BOY NAMED PLUTO

INTRODUCTION

Once upon a time, in a neighborhood far, far away, there was an unusual group of friends who lived together on a street called Milky Way. This group of neighbors had known each other for a very long time – sometimes they even joked that it seemed like a million years. They have always been relatively close friends, but in many ways, they circled about in their independent lives with many unique characteristics. However, no matter how different they were, they all still lived happily together in their peaceful neighborhood.

But all of this changed on that day in 2006; the day that IAU made the announcement.

A BOY NAMED PLUTO

CHAPTER 1

The Bad News

E ach of their houses was scattered down Milky Way, extending about a half-mile from the beginning to the end. One time they all happened to be getting their mail at the same time and they all waved and shouted to one another, teasing about how they were dressed, or about the weeds in one yard, or asking about the kids, or discussing things that neighbors chat about.

On the day the news came out, it spread like a comet. Of

course, Pluto, who lived at the end of the street, was the last one to find out.

Everyone has a deep affection for Pluto, and they keep a watchful eye out for their little companion. After all, he is only eight years old and lives all alone at the end of the street. Someone told him that when he was younger, his parents got into a huge fight and in an instant, his mom stormed out the front door and his dad stormed out the back, both slamming the door in one big bang – never to return.

From Pluto's place, there are no houses, farms, or buildings anywhere in sight. The street in front of his house goes straight off into the distance to a small spec at the horizon. Fortunately, his perfect companion, Nix, is always there to keep him company day and night so he doesn't get too lonely.

For most of his life, Pluto has felt like he could never quite fit in with the others on Milky Way. *Maybe it's because I live so far away*, he thinks.

But that never stops him from having a wild imagination. Like most eight-year-olds, Pluto is full of energy and loves playing make-believe. He often thinks about far-off places and he dreams. He dreams he's a dinosaur roaming the planet eating tall trees and crushing buildings under his feet. He dreams he is a fireman, saving families from burning buildings. He dreams he is a space traveler riding a spaceship throughout the universe exploring planets and galaxies with unusual names. Most of all, he dreams of being a grown-up just like all of his friends on his street that he so admires. "Soon, little buddy" they would say. And he would wait patiently, even though he had no idea what 'soon' really meant.

Pluto's favorite spot in the whole city is his backyard where he can see a very faint cloud, low on the horizon, extending around the entire skyline. He loves to watch this cloud, which seems to spin slowly in a counterclockwise direction around the sky. He wonders what is out there and he imagines riding his spaceship to this cloud one day to see what is there.

The one thing that Pluto enjoys the most is reading. Even though he is young, he has the reading skills of someone much older. This ability allows him to submerge himself into books of all types for hours and hours at a time. But his favorite thing to read is the Daily Globe morning paper.

Each day with very little warning, Rocky the paperboy will come flying down the street at such a speed, you can hardly see Rocky's odd shaped and hardened face. He flings the paper, bundled up in a plastic bag, onto Pluto's porch at such force and velocity that when it hits the front siding of the house, it makes little pock marks in the soft siding. Over the years, the front face of the house has gotten so pelted with dents and craters that it looks like the surface of the moon.

But Pluto doesn't care. He loves reading every single square inch of every section. It is his way to keep connected to what is going on in the universe.

That was when he saw it: On page 4, Section D, Column 3 at the very bottom. It was a little afterthought dropped in by the editor because there was some extra space the editor needed to fill. The headline read:

ASTRONOMERS DECLARE PLUTO NO LONGER A PLANET.

HEARING AND FINAL DECISION IN 3 DAYS.

Pluto was shocked.

A BOY NAMED PLUTO

CHAPTER 2

Getting Help

Tooner came flying up the front porch and completely ignored Nix who was overjoyed to see his buddy. Tooner ran straight into Pluto's house just as Pluto finished reading the article.

"Can you believe it!?" exclaimed Tooner. "It can't be! This is really bad news."

Tooner had ridden over to Pluto's house on his light grey Razor with the lightning bolt decal on the top, half worn-out near the back wheel from Tooner dragging the brake when taking curves. He always rides his scooter recklessly around town. In fact, one day he crashed into the mailbox of old man Earl who lives six houses down on Milky Way and got a nasty bruise on his hip that has never gone away.

Tooner tried his best to comfort Pluto, but he could see that Pluto was very disappointed and sad.

"This is so lame!" Tooner shouted – not to Pluto in particular, but to anyone that happened to be around, even though Tooner knew they were alone. "You don't deserve this."

They had no idea what to do about what they had just read in the paper. But they did know that Ursula could help.

So off they went to Ursula's house with Nix bouncing down the road and Pluto running alongside Tooner in order to keep up.

It was a nice day out and as they went along, Tooner tried to chat about various things to help keep Pluto's mind distracted. Tooner brought up how both of them had a fondness of cold weather. They talked about how they periodically play outside, in the dark, in the cold, for hours and hours at a time and how fun it is to do that when no one is around. As Tooner brushed away a circling fly in front of his face, he talked about the time when they laid on their backs in the cool dirt and watched shooting stars light up the sky. Pluto played along with the conversation, but Tooner could tell his mind was somewhere else so he picked up the pace to Ursula's house.

Neither one of them noticed a black 4 door Lincoln Town Car parked down the street about a hundred yards away. And certainly, they didn't see the long black telescope poking out the rear rider side window watching their every move. No strangers ever visited Milky Way. Who could this be?

As they approached Ursula's house, they yelled out from the sidewalk as soon as they knew Ursula could hear them.

"Ursula!!" They both shouted in unison. "We need you!"

Ursula, wearing her favorite light-blue sun dress and sitting on her couch, was knitting a blanket for her upcoming first grandchild in the fall. She was focused on the cross stitching technique to create a yellow and blue duckling in each corner. Ursula lived alone after her husband died a decade ago and she really enjoyed visitors, especially energetic young Tooner, even if he consistently didn't give any notice before stopping by. Fortunately, Ursula accepted both Tooner and the flies, just as long as they wouldn't go near her cookies and donuts in the kitchen.

Without so much as a knock, Pluto and Tooner bolted into the living room with Nix trailing behind. Ursula wasn't startled in the least by their entrance, as she was expecting them. She had read the article earlier and knew how devastating it would be for poor little Pluto.

"Pluto, my dear," she said in a soft comforting voice, looking up from her bifocals.

"Come here, little one." She arose from the couch to give Pluto a hug. "We will get you through this. Ever since I read the article, my mind has been spinning in all directions, trying to comprehend what this is all about. I may not know exactly *what* you should do, but I do know *who* you need to talk to."

Ursula reached over for the paper on the coffee table to glance at the article one more time.

"Vinny has always been the brightest one in the whole neighborhood," she continued. "He'll know exactly what to do to get this fixed! Since you only have 3 days, I think we need to head over to see him right away!"

At least now Pluto felt a bit of hope that Vinny might know what to do.

"It's a bit of a walk to Vinny's so let me get my coat," Ursula said. "You know how cold I always get."

Ursula came back with her gray coat and white sneakers and the three of them headed out the door, but not before Tooner grabbed a few cookies for the trip.

CHAPTER 3

Heading to Vinny's

R inger was sitting on his front steps eating a sausage and pepperoni pizza. Despite his parents constantly telling him otherwise, he was blasting his favorite band through his headphones from his Galaxy GS7 smartphone. When he heard about Pluto earlier that day, all of his rings, pierced and magnetic earrings and body piercings turned shades of dark blue because he felt bad. Real bad.

He was minding his own business when all of a sudden, Nix ran up and snagged Ringer's slice of pizza right out of his hands, taking off running down the street.

"Hey, you little rascal!" screamed Ringer. "Bring that back!"

Within seconds, Ringer's mood rings turned red as he darted off to catch Nix. By the time Ringer caught up with Nix, he was hiding behind Pluto's legs.

"Oh no, I'm sorry Ringer!" Pluto exclaimed as Nix tried to hide his prize. "Nix is normally better behaved."

"That's OK," says Ringer. "By the way, sorry about the news, pal."

Ringer realized that he couldn't stay mad for long and his rings turned back to a soothing light blue. "You deserve to be a planet and it's just not right what they are doing."

"Thank you, Ringer," Pluto replies, a little embarrassed that his dog just ate his neighbor's pizza.

"I really want to help you guys," Ringer comments, his words slightly intermixed with a little bit of jingling from the rings hitting together on his lips. "Where are you going?"

"We are heading to Vinny's because he'll know what to do!" Ursula explains.

Without a second of hesitation, Ringer says, "I'm in, dudes," and shoots to his steps to get the rest of the pizza then heads back to the street to join the trek.

The group starts hustling down the street toward Vinny's and it wasn't long before Jesse comes jogging up.

"What's uuuuup?" Jesse asks the group in a slightly winded voice while pulling up behind them in his headband, pulse meter, and tight nylon pants that accentuate his larger-than-normal calves.

"We are heading to Vinny's house to get help for Pluto," Tooner explains.

"Oh yeah, I forgot about that, Pluto," Jesse comments. "Sorry to hear, man. I was planning to come over to see you today but I had to get my workout in first, you know. I'm not sure what Vinny thinks, but if I get my hands on the guy that made this decision, I'll put him in a headlock until his eyes pop out and he changes his mind."

Jesse snarled his teeth while forming his right arm into a 'squeezing V' to make his bicep bulge as big as possible.

"Mind if I tag along?" he asks.

Having some brawn might come in handy, Ursula thought to herself.

"Sure, you can join us but we have to hurry. The paper said we only have 3 days!" she replied.

Pluto, Tooner, Ursula, Ringer, Jesse and Nix advance down the street: Jesse jogging-in-place but moving forward to keep pace with the others, Tooner riding his Razor pushing with one foot and gliding with the other, Nix sniffing from mailbox to mailbox, Ringer zoning out in his out-of-this-world punk rock music, and Ursula and Pluto keeping up while holding hands at the back of the pack.

Catching them by surprise, the black Lincoln suddenly zipped by them. Inside, were two men. They were dressed identically in black suits with black fedora hats pulled down low. Both of them wore white shirts, narrow black ties, and dark sunglasses with opaque lenses. It appeared as though they came from *Men in Black*, one of Ringer's favorite movies.

As the Lincoln drove by, both men slowly stared at the group, turning their heads in unison to watch them as the car passed by.

No one knew who these strangers were, but they were glad to see them leave. The car and the men made everyone uncomfortable for some reason, especially in a neighborhood as pleasant as theirs.

As the car passed on ahead of them, not one member of the group noticed the bumper sticker on the back left of the car. In black lettering with a white background, surrounded by a thick black oval, were three letters: **IAU**.

A BOY NAMED PLUTO

CHAPTER 4

Marti Joins In

As the group continued toward Vinny's house, they approached Marti's single-story rambler with the red tiles on the roof and light tan stucco siding.

Marti is a beautiful red-head who loves the outdoors. As a hiker and mountain climber, she is part of an explorer club that climbs some of the tallest mountains on the planet. In fact, some of her fellow explorers, such as the Minnesotan she nicknamed the Viking Lander and the Pathfinder who can map out the best climbing routes for any mountain, are famous for leading manned and unmanned explorations all over the world.

Marti, wearing tan shorts, a red plaid flannel shirt, and hiking boots, was out on her lawn setting up a tent that needed to dry out in the sun.

"Whatcha doing, boys? Marti shouted over to the group on the street.

"We're heading to Vinny's to get help for Pluto," Turner replied. "Want to come?"

"Absolutely! I've been thinking about how I could possibly help." Let me change quickly."

So Marti ran into her house and returned wearing a red sundress and several metallic necklaces that have the appearance of being rusty, yet beautiful. But she kept her warm hiking boots on because she never left the house without making sure her cold feet were covered.

"I sure hope Vinny can help," Ursula nervously says to the group.

"Me too." chimes in Pluto.

A BOY NAMED PLUTO

CHAPTER 5

Earl's Business Partners

Vinny's house was just a short walk from Marti's and on the way, the gang passed Earl's home, and he was out watering the flower garden. As the group walked by, they waved to Earl and he gave a welcomed wave back, obviously wishing he was somewhere other than standing in a garden with a hose. His wife was out of town visiting her sister so he was given the task of watering the roses while she was gone. Nevertheless, he shouted out "Have a great day!" as if he was a teacher when the bell rang and the students headed out the classroom door on a Friday afternoon.

He must not have heard the news. Otherwise I bet he would have said something to Pluto, thinks Ringer.

Pluto and the gang keep their focus and march ahead to Vinny's as quickly as possible.

Tooner wasn't paying attention and he got too close to Jesse in front of him and he accidently rammed his Razor up the back of Jesse's tennis shoe causing Jesse's heel to come out of the shoe.

"Hey, cut it out!" Jesse snapped, giving Tooner a sharp look. "Don't be so dense!" Even though Tooner wasn't the smartest of the group, it wasn't very nice of Jesse to say that.

As Jesse bent over to hook his shoe back onto his foot, he looked back at Earl's house. At that very moment, the black Lincoln was parked in Earl's driveway and the two guys in black suits were talking with Earl. "Hey, guys! Look!" he shouted and pointed to Earl. They all turned around and saw Earl shaking hands with each of the men. Then Earl got into the back seat and they took off in a hurry.

"Why would Earl be talking with those men?" Pluto asked Ursula in a soft innocent voice.

"It does seem to be a bit confusing, but it'll be ok. Come on everyone. Quit thinking about Earl. We have to get to Vinny's. It's just up ahead."

"You think Vinny will have anything to eat?" Ringer inquires.

Tooner swats a fly, misses in his attempt, then comments quietly "Just *once* I would like to catch one."

A BOY NAMED PLUTO

CHAPTER 6

Vinny's Plan

Vinny was hanging out in his living room with his next door neighbor, Merk, who had FaceTimed him earlier when the paper came out. Merk was sitting on the couch next to Vinny eating Flaming Hot Cheetos and wearing his sunglasses even though he was inside. Merk loves the sun and heat so much that he doesn't even realize he's wearing them.

Merk got his name because his high school buddies razzed

him about driving a beat-up Mercury Mariner SUV that his folks gave him when all his friends had little sports cars, Honda Accords, and jacked-up pick-up trucks. But he was one cool cat.

Knock! Knock! Knock!

"Hey, Vinny – You home?" Jesse shouts from outside.

The door swings open and Vinny anxiously scans over the group.

"I'm so glad you're here," he said. "There's not much time! Come in."

Marti was the first to enter, followed by the others.

"Hey, Vinny. Hi, Merk." Marti exclaims as she lightly bounces into Vinny's house with her ponytail bobbing behind her neck. "What ya doing?"

Vinny cuts right to the chase and explains to the group that the courthouse is making their declaration in only 3 hours.

"Three *hours*?" Ursula asks. "The paper said it was going to happen in three *days*".

"It took me a bit longer than normal to read this, but according to the internet, three days here is like three hours there!" Vinny says, pushing his glasses up his nose. "Nevertheless, my computer indicates that we have to get to the hearing right away and convince them that Pluto is really a planet."

"Convince who?" asked Pluto, with a look of worry and confusion on his face.

"The International Astronomical Union," answers Vinny. "Take a look at this."

Vinny gathers the group over to his computer screen and shows them what he's found.

"When I read the announcement in the paper this morning, I began researching what happened. It seems that there is this group of astronomers that analyze the whole universe and determine who is who and what is what. They claim that Pluto doesn't fit the criteria as a planet. They say he's a *dwarf* planet."

"That's not fair!" Ringer interrupts. "He's just the same as us!"

"Yeah, and I don't even know what a dwarf planet is!" Tooner exclaims.

Pluto looks up at Ursula with longing eyes, tugs on her dress, and gently tells her in a childlike voice, "I don't want to be a dwarf. I want to grow up to be a real-sized planet just like all of you."

"I have an idea," says Vinny. "It says here that the hearing is open to the public, so we have a chance to change their minds if we can just get to the courthouse. There's no time so we have to hurry! Can we use your car, Merk?"

"That's cool with me," Merk responds as he licks his fingers, trying to remove the orange stains from the Cheetos. "Can someone else, drive, though?" He silently hopes everyone will be okay with the fact that he removed his car's air conditioning years ago.

"Okay great. Huddle up. I have a plan," Vinny states.

Eight of them, plus Nix, pile into Merk's car and Jesse starts backing out of the driveway. "Wait a minute!" shouts Vinny. "I forgot something! Hold on." He jumps out the back of the car and runs into the house. Less than a minute later, he comes bolting out and nearly trips coming down the steps.

"What was that all about?" Ringer asked.

"I had to do something on the computer. You'll see." Vinny hops in the back, closing the car door behind him. "Burn it up, Jesse. We don't want to be late."

That's all Jesse needed to hear. He tore out of the driveway and flew down the road like a rocket.

CHAPTER 7

The Courtroom

They made it – with a minute to spare. And they were lucky to have a parking spot open right in front of the courthouse next to the handicapped parking spot – and with time on the parking meter to boot. They hoped their good luck would continue when they got inside.

The courtroom was full of energy. The decision to declassify Pluto's planet status apparently caused quite a buzz near and far.

There were reporters from magazines and newspapers. There were officials from NASA. And there was even a group of scientists from Russia who studied in the USA before joining the Roscosmos, Russia's counterpart to NASA.

Ursula and Marti entered the courtroom first and took a seat in the second row. They glanced around the room and saw the elevated judge's bench towering down above everything else in the entire room. It was made out of hand-carved oak, stained in a dark red and it portrayed power and control, not to mention how intimidating it was to the average person.

Ringer and Tooner came in next. Immediately behind them came Jesse and Merk. Merk still wore his tropical shirt, confidently and coolly strolling in; a major contrast to his walk mate, Jesse, who bounced in all cocky and proud.

All four sat in the front row, directly facing the judge's bench. In front of them was a table with two chairs – the kind of chairs which were supposed to be comfortable, but in reality, were stiff and undersized, specifically designed to keep the people sitting in them from staying too long. Off to the right, was a court recorder seated at a small table.

Then they saw them. The men from the black Lincoln. Both of them were sitting on the other side of the room, facing forward and not moving an inch. In front of the men was an opened briefcase, waiting patiently to be tapped into.

"All rise!" was bellowed out by someone in the back with a deep, penetrating voice that felt like it would hurt you all by itself.

Immediately, a large oak-paneled door behind the judge's bench opened slowly and in walked a large intimidating man with a black robe and white wavy hair. He was looking down reading papers as he stepped up to the seat behind the bench.

What was already a dire situation suddenly turned worse. When the judge pounded the gavel to begin the hearing, every single eye was on him – and what a shock it was when the judge looked up and stared directly at them. IT WAS EARL! And Earl clearly was friends with the bad men from the black Lincoln! What else could go wrong? They hadn't even started the hearing yet and they already had two strikes against them.

CHAPTER 8

Support for Pluto

As intimidating as the whole situation was, the friends did have something good coming their way that they weren't aware of yet: All of the energy that they felt in the room wasn't just imagined. It was real. The entire crowd was there to help Pluto!

Tooner, Ursula, Merk, Jesse, and Marti were completely overwhelmed by who had showed up to support their beloved neighbor.

There were supporters from all over the town. Halley's Comet, Rocky with his fellow asteroid buddies, stars and movie stars. The moon was holding a "Save Pluto" sign. The Sun was cheering away. There was even a Black Hole that was off in a corner (by himself, fortunately).

More cosmic friends were in the back cheering: Meteors and meteorites were holding hands and swaying back and forth, several galaxies were wearing **I ♥ Pluto** T-Shirts. The Big Dipper and Orion were holding up lighters. The solar system was shouting out "Good Luck Pluto!"

All sorts of spacecraft and agency apparatus had flooded the courtroom wearing armbands or hats in protest: the Lunar Lander, the International Space Station, the Mir, satellites of all kinds, the Endeavor and Atlantis Space Shuttles, Sputnik, twin brothers Voyager I and II, and much more.

Even a host of astronauts showed up wearing Pluto patches on their uniforms: Neil Armstrong, Sally Ride, Yuri Gagarin, John

Glenn, and Alan Sheppard to name but a few.

Everyone started chanting "We want Pluto!" followed by the five-clap repetition like it was a high school basketball game.

The solar Tweets that Vinny sent out at the last minute worked! Everyone joined in to lend their support to help Pluto remain classified as a planet. It was astonishing.

Then it happened. The big front double doors of the courtroom swung open and in came Pluto with Vinny trailing slightly behind him. The courtroom went silent.

As Pluto proceeded down the aisle to the front table, he had been transformed. With the help of his neighbors, he walked with confidence and portrayed himself as an individual much beyond his eight years of age.

He now sported a Universe Wrestling Foundation belt. He was wearing dark sunglasses. He had rusty jewelry around his neck. He had an earring in his right ear. He was wearing a flowered sweater, and he was gliding along on a Razor. All of these things, he believed, would show the court that he was like the other planets and deserved to be one himself.

He and Vinny sat down at the front table and the hearing began.

CHAPTER 9

The Arguments

The men of IAU presented first. They had one central argument to their position: Eris.

"We would like to present Eris as evidence to this court," one of the men announced. "Eris, please stand for everyone to see."

Out of nowhere, a short girl with long blonde hair and an angelic face arose from the crowd and faced the judge's bench.

"You see, Judge, Eris lives in a distant place called Kuiper, the next city down the road past Pluto's house. Officially, Eris is a dwarf planet. But to most, she looks just like a little girl you see playing in the park. The entire city of Kuiper is comprised of dwarf planets like Eris. Kuiper circles around the horizon in a large continuous group off in the distance. Kuiper is actually visible from Pluto's backyard."

The IAU men then proceeded to explain all of the reasons why Pluto is more like Eris than he is like Tooner, or Jesse, or Ursula or Pluto's other friends. They presented all kinds of technical information to support their position, including data and results from astrophysicists and expensive analysis equipment with names that are hard to pronounce and even harder to spell.

"It's possible that Pluto was cast-off from Kuiper by mistake," they reasoned. "Maybe he accidentally was caught in the gravitational pull of the Sun when he got isolated from the Kuiper belt. All that it would require is to wander out to pick up a lost toy or to slip on ice while being curious about something flying by – and he got separated."

But, for the most part, they concentrated on one key fact.

"To be a planet, you need to A) be shaped like a ball, B) circle around the sun, and C) when you go around the sun, you can't cross anyone else's path. We have been observing Pluto for some time and we do agree that Pluto is round in shape and that he might take a long walk around the sun once in a while. However, he fails to comply with the third requirement."

The men pulled out what appeared to be a recording device from their telescope and projected videos and photos onto a screen at the front of the room.

"We have personally witnessed and recorded multiple instances of Pluto crossing paths with Tooner," continued the IAU.

Pluto and Tooner glanced over at each other from across the table. Neither of them had any idea that they were being spied on in the first place.

"Not only that, you can see in these photos that Pluto also regularly ventures into Kuiper city limits."

"I didn't know I was even crossing the limits," Pluto whispered to Vinny.

"The law clearly states that a *planet* must meet all three criteria. Therefore, since we have evidence that Pluto does not meet all three, he is to be classified as a dwarf planet. There is just no other way to conclude otherwise."

The room remained silent as the IAU took their seats.

"Alright, I've heard enough," Earl interjects. "I must say, you gentlemen bring up some very valid points. I'm quite sure there is not much more to discuss here."

But Vinny had his own argument.

When it was his turn, Vinny asked Judge Earl if it was okay to have Pluto testify on his own behalf.

"Go right ahead," Earl said with a devilish grin. "Although, I don't really see the point."

Pluto arose from his seat and slowly made his way to the

witness stand next to the Judge's bench. Vinny began his case on why Pluto is a planet.

"The IAU certainly has the wonderful scientific equipment to support their position," Vinny began. "Well, two of their own high-powered devices will prove that Pluto is a planet and should remain a planet."

Vinny pulled out an 8" X 10" full-color photograph of Pluto. "I present this photograph to the court that is taken by none other than the world-famous Hubble Telescope. You can see clearly in this photo that Pluto looks just like the other planets. The only difference is just that he lives so far away."

Vinny then pulled out a dozen photographs and held them up high with his left hand while pointing to the sky with the index finger on his right hand.

"These photographs present even more evidence of that Pluto is like a planet," continued Vinny. "These were taken less than one month ago by the space probe, New Horizons, which took thousands of pictures of Pluto: zoomed in, up-close and from far away - all of them proving he is like the other planets."

Vinny then hands the photographs over to Pluto.

"My friend, here are pictures of you from the IAU's own equipment and it presents our case that you are indeed a planet. At this point, I would like to hear from you. Tell me what you feel about these photographs."

Vinny didn't know it, but he had just changed everything.

Pluto took the pictures and stared at them for a very long time. The entire courtroom was silent as they watched little Pluto study all the photos.

He glanced up to look at his friends, he looked over at Eris, and then Pluto looked at the photos once again.

At that point, he began to slowly speak. "Is this really how I look?" Taking a long breath, he says, "I have never seen myself before."

Pluto then paused, trying hard to hold back tears. "Everything I've ever known has been because of my friends. All I've ever wanted to be when I grew up was a planet, just like them."

He nodded his head down slightly and started to sniffle.

"But I… I… look… like….you," Pluto hesitantly exclaims, slowly pointing with his right hand to the little girl sitting behind the men from the IAU.

At that very moment, Eris rushed up to Pluto and gently touched the back of his shaking hand.

"It's alright, Pluto," she whispered in her kindest little girl voice. "You *are* like me. And you are <u>*just like all my friends*</u>. It's okay."

Then Pluto reached down and slowly removed his UWF belt, the rusty jewelry, and all of the extras that his pals gave him. He placed them carefully on the rail in front of him.

There was nothing more to do. But everyone knew what to do next.

Chapter 10

All Is Good

The courtroom started clapping and cheering – a little bit at first, then it grew. Everyone was thrilled that the issue was settled and that Pluto was happy.

First Ursula went up to Pluto, followed by the rest of the crew. And then others came up. Everyone in the entire courtroom instantly became sympathetic and realized that it really didn't matter that he was a dwarf planet or not. Pluto would still be their beloved friend no matter what.

Fairly quickly, Pluto was surrounded by the crowd that had come to support him, giving him hugs and high-fives, and eventually, he was hauled up onto the shoulders of Jesse. As he was carried outside, there was small little Pluto, perched up high and above everyone in the whole universe. He had a grin from ear-to-ear as the crowd followed him out of the courtroom. Even Earl had come down off the bench and joined the crowd as they proceeded outside, chanting and cheering Pluto's name as they marched.

"Look how happy he is to have found his place," Marti whispered to Merk, as they made their way to the exit.

"Totally," Merk agreed. "He can now play and interact every day with people just like him."

Scooter, Ringer and the rest of the gang were waiting outside for Pluto as he said his goodbyes. No longer did they feel sorry for him living so far away on Milky Way. No longer were they concerned that he might be lonely and out of communication. Now, all of his planet friends were comforted by the fact that Pluto would finally fit in.

Pluto gave everyone a hug, one after another; a little, but very impactful squeeze that will never be forgotten by each of his friends. The last to get a hug was Ursula.

"I'm going to miss you the most," Pluto tells Ursula in the sweetest little boy voice she has ever heard. His hug is extra-long as Pluto's arms start to shake a bit as he squeezes around her neck.

"I know, honey," Ursula responds, talking gently with her tilted head lying on top of his buzz-cut hair as he laid against her shoulder. "But you'll do fine. Sometimes you go about thinking one thing and something completely different happens. Even though we'll be close by, you will be much happier being with friends and others like you. Just don't forget about us – we will always remember you."

The embrace ended and Ursula pulled out a crinkled up, semi-used Kleenex from the front pocket of her dress and gently wiped her nose. "We love you."

As he turned away from Ursula, Pluto looked over and saw Eris standing near a tree in her pink and white dress with a matching bow in her hair. He walked slowly and hesitantly over to Eris to leave with her. Glancing over his shoulder from a distance, he looked back at the group and waved slowly with a little curled-hand type of wave.

"We'll miss you, little buddy," Tooner whispers to himself as he half-heartedly swipes at another circling fly, tears rolling down his cheek. He sniffled but smiled, watching in the distance as Eris and Pluto walk away holding hands, with Nix bouncing around them.

<The End>

A BOY NAMED PLUTO

DID YOU CATCH THE CONNECTION?

A Boy Named Pluto is full of analogies and factual references that you may find interesting. Check out these suprising facts!

Pluto
- Pluto was originally named planet "X" when discovered in 1930 by the Lowell Observatory in Flagstaff, AZ.
- An 11-year old girl named Venetia Burney from England suggested the name Pluto.
- Pluto's parents leaving in a 'big bang' represents the name that scientists call the event at the origin of the universe (pg 7).
- Pluto is a very small planet approximately 1/3rd the size of the moon.
- The IAU announced on May 14th, 2006 that Pluto would be classified as a dwarf planet (pg 10).

Tooner
- Tooner represents Neptune, which has an interior composed of rock and ice, thus getting the nickname as an Ice Giant.
- Tooner's bruise on his hip represents the Great Dark Spot in Neptune's mid-section (pg 13).
- Pluto and Tooner playing together in the cold is in direct reference to the pair of adjacent planets that are more than 2.7 billion miles away from the sun. Both planets are colder than -350F (pg 14).
- Neptune is the most dense of the planets as Jesse mentions walking to Vinny's house (pg 29).

Ursula
- Uranus is the coldest planet (as low as -224C) and it is very windy with wind speeds reaching 560 mph. Therefore, Ursula needs to get a coat for the journey (pg 15).
- Due to its methane-based upper atmosphere which absorbs red light waves, Ursula wears blue to signify the blue tint appearance of Uranus (pg 15).
- The planet's long 84 year orbit is the basis for Ursula living alone (pg 15).
- Ursula's comment about her head spinning is due to Uranus having an inverted axis where the planet rotates top to bottom vs. side to side (pg 16).

Ringer
- The rings on Ringer's hands and his earrings represent Saturn's spectacular rings that are made up of countless particles made of water and ice with a trace of rocky material (pg 18).
- Ringer's bipolar disorder is reflected in the changing colors of Saturn's rings and Ringer's mood rings (pg 18).
- Ringer's magnetic earring represents Saturn's magnetic poles which are similar to Earth's, but slightly weaker (pg 18).
- Saturn has a large moon called Titan and is the only moon with its own atmosphere.

A BOY NAMED PLUTO

Jesse
- As a body builder, Jesse represents Jupiter, the largest planet at 88,000 miles across (pg 21).
- Jupiter completes a full rotation in only 9.8 hours therefore Jesse is portrayed as a fast runner (pg 20).
- Jesse's bulging muscles represent Jupiter's bulging equator (pg 21).
- Jupiter's 'Geat Red Spot' near its equator and its horizontal stripes are presented as Jesse's UWF belt (pg 20).
- Jesse being full of hot-air ties back to Jupiter being known as a gas giant, with most of its atmosphere comprised of hydrogen and helium (pg 20).
- Vinny telling Jesse to "Burn it Up" is a reference to Jupiter's temperature extremes which can have areas of its upper atmosphere reach up to 1,340 degrees F (pg 36).

Marti
- The tallest mountain in the solar system is called Olympus Mons located on Mars and is why Marti is a mountain climber. It is twice the height of Mount Everest. (pg 24).
- Mars is the planet that has experienced the most exploration by scientists with more than 2 dozen unmanned spacecraft (pg 25).
- Marti (a redhead and her attraction to the color red) correlates with Mars being called the "Red" planet (pg 24).
- The color of the planet is due to iron oxide (rust) on rocks on the surface which is presented as Marti's rusty necklace (pg 25).
- Marti's cold feet/boots and the white tuft of hair represent the planet's polar caps (pg 25).

Earl
- Representing Earth, Earl's round face plus his white, wavy, and changing hair is an analogy for the planet's ever-changing clouds and weather.
- Earl watering the garden presents to readers that the earth is the only planet with an abundance of water-supporting life (pg 28).

Vinny
- Vinny represents Venus, the planet with the longest solar orbit of any planet in the solar system (243 earth days).
- Venus is the brightest planet in the nighttime sky therefore Vinny is portrayed as the brightest intellectually (pg 16).
- Vinny is dyslexic, directly correlating to the fact that Venus rotates on its axis in the opposite direction than other planets (pg 35).

Merk
- Mercury, cast as Merk, is the closest planet to the sun (29 million miles from the sun and 48 million miles from the Earth).
- Presented as a character loving the sun and heat, Mercury is the hottest planet in the solar system. Temperatures on Mercury can reach 800F (pg 32).
- With Merk loving the heat, but portrayed as 'cool', presents to readers that Mercury has a wide temperature range, getting very hot in the daytime, but also very cold (-279F) at night (pg 32).

A BOY NAMED PLUTO

Eris
- Eris (the little girl) represents a dwarf planet also named Eris that resides in the Kuiper Belt. Eris has one known moon (pg 46).
- Eris is the second largest and the second most massive dwarf planet behind Pluto.
- Eris is the largest known body in the Solar System that has not been visited by a spacecraft.

IAU
- The International Astronomical Union comprised of more than 12,000 professional astronomers from 98 countries all over the world, announced on May 14th, 2006 that Pluto would be classified as a dwarf planet (pg 23).

Nix

- Nix is the name of the second largest of Pluto's five moons and is represented by Pluto's dog (pg 8). It was discovered in 2005 using the Hubble Telescope (pg 7).
- Nix the moon circles and rotates about continuously in various directions as does the dog in the story (pg 13).

Black Hole
- A black hole is a place in space where gravity pulls so much that even light cannot get out. The gravity is so strong because matter has been squeezed into a tiny space. This can happen when a star is dying. Because no light can get out, people can't see black holes. They are invisible (pg 43).

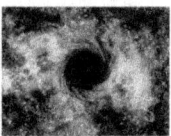

The Viking Lander
- NASA's Viking Projects (1 & 2) were the first U.S. missions to land a spacecraft safely on the surface of Mars and return images of the surface in 1976 (pg 25).

The Pathfinder Probe
- In 1997, the Pathfinder Space Probe utilized an air-inflated bag system to successfully land on the surface of Mars (pg 25).

Rocky the Paperboy
- Rocky is an asteroid, which is made of hardened rock in irregular shapes. Asteroids can range in size from a dust particle to 600 miles across. Rocky's speed of delivery represents the fact that Asteroids can fly more than 50,000 mph (pg 9).

Meteors and Meteorites
- A meteor creates a bright trail or streak of light in the night sky when it enters the Earth's atmosphere. The friction with the air causes the rock to glow with heat.
- Also called shooting star, most meteors burn up before reaching the Earth.
- A meteorite is a meteor that successfully passes through the earth's atmosphere and part of it strikes the ground (pg 43).

Halley's Comet
- Likely the most famous comet, Halley's Comet flies near earth every 75 years, returning next in 2061 (pg 43).

A BOY NAMED PLUTO

The Big Dipper
- A group of seven stars within the constellation formally called Ursa Major (the Great Bear) (pg 62).
- As a subset of a constellation, the Big Dipper is officially an asterism.
- The term Big Dipper was used because the 7 stars form what looks like a large spoon/soup ladle (dipper).
- The two bottom stars point to the North Star.

Orion
- A constellation representing a human hunter and is recognizable due to three bright stars making up a perceived belt. Other stars comprise the shoulders, legs, and arms (pg 43).

Galaxy
- A system of millions or billions of stars, together with gas and dust, held together by gravitational attraction (pg 18).

The Lunar Lander
- Famous landing craft of Apollo 11 space mission carrying the first men to land on the moon on July 19th, 1969 (pg 43).

The International Space Station
- The International Space Station (ISS) is a space station and artificial satellite that is in low Earth orbit. Its first component was launched into orbit in 1998.
- The ISS is now the largest artificial body in orbit and it can often be seen with the naked eye from Earth (pg 43).

Endeavor and Atlantis
- Space shuttle Endeavour is a retired orbiter from NASA's program and the fifth and final operational shuttle built. It embarked on its first mission in May 1992 and its 25th and final mission, in May 2011 (pg 43).
- Atlantis is the fourth operational and the second-to-last space shuttle built. Its maiden flight was in October 1985 while its 33rd and final mission, also the final mission of a space shuttle, was in July 2011.

Mir
- Mir was a Soviet Union space station that operated in low Earth orbit from 1986 to 2001. Mir was the first modular space station and was assembled in orbit from 1986 to 1996. Until March 2001 it was the largest artificial satellite in orbit (pg 43).

A BOY NAMED PLUTO

Voyager
- Voyager 1 and 2 are space probes launched by NASA in September 1977 to study the outer Solar System (pg 43).

Sputnik
- Sputnik was the world's first artificial Earth satellite launched by the Soviet Union into an elliptical low Earth orbit on 4 October 1957. It was a 58 cm (23 inch) diameter polished metal sphere, with four external radio antennae and weighed 83.6 kg or 183.9 pounds.
- The 98 minute mission initiated the space race, ushering in new political, military, technological, and scientific developments (pg 43).

The Milky Way
- Milky Way is the galaxy where Earth resides, but in this story, the street name "Milky Way" is an analogy for our solar system.
- The houses of each character on the street called Milky Way are presented in the same order as the planets in the solar system with Merk living closest to the sun and Pluto living the farthest from the sun (pg 4).
- Scientists estimate that the Milky Way and the entire universe were formed 13.7 billion years ago.

New Horizons
- New Horizons is an interplanetary space probe that was launched as a part of NASA's New Frontiers program. The spacecraft was launched in 2006 with the primary mission to perform a flyby study of Pluto and a secondary mission to study one or more other Kuiper Belt objects. It completed its study of Pluto in 2015 (pg 50).

Hubble Telescope
- The Hubble Space Telescope is a space telescope that was launched into low Earth orbit in 1990 and is one of the largest and most versatile. It is well known as both a vital research tool and a public relations boon for astronomy.
- The telescope is named after the astronomer Edwin Hubble (pg 50).

Solar Flare
- A sudden eruption of magnetic energy released on or near the surface of the sun, usually associated with sunspots and accompanied by bursts of electromagnetic radiation and particles (pg 44). The solar 'Tweets" sent by Vinny represent solar flares.

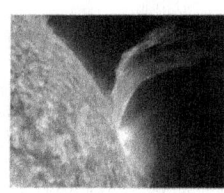

A BOY NAMED PLUTO

Neil Armstrong
- Armstrong was the first man to set foot on the moon on July 21st, 1969 from the Apollo 11 Lunar Lander. He was famous for stating, "That's one small step for man, one giant leap for mankind."

Sally Ride
- Ride was the first American woman to fly in space in 1983.

John Glenn
- Glenn was the first American to orbit the earth in 1962.

Yuri Gagarin
- First human to journey into outer space, when his spacecraft completed an orbit of the Earth on 12 April 1961.

Alan Sheppard
- Became the second human and first American in space with the Mercury flight in May of 1961.

OTHER INFORMATION

Planets vs. Dwarf Planets (pg 48)
- According to The IAU, a planet must have three criteria while a dwarf planet has only two.
  1. It is in orbit around the sun.
  2. It has sufficient mass for its own gravity to pull itself into a nearly round shape.
  3. It has not cleared the 'neighborhood' surrounding its orbit.

Pluto does not meet #3 due to the fact that its orbit intercepts with both Neptune's orbit and the Kuiper Belt.

Kuiper Belt (pg 46)
- The Kuiper Belt is a disc-shaped region of icy bodies located outside the orbit of Pluto. The belt includes dwarf planets such as Pluto and Eris, hundreds of thousands of icy bodies, and an estimated trillion or more comets.

The 3 Day Quest (pg 34)
- The reference of the timeframe between three days and three hours is an intentional inference to show that the each planet has differing lengths of days (rotational speed) and varying length of years (rotation around the sun).

ABOUT THE AUTHORS

The father/daughter team of Ashley Gallegos and Jahn Stopperan have jointly blended their creative writing abilities in forming this book and have crossed generations in presenting an imaginative story with an educational theme.

The authors are business marketing professionals who have significant experience writing technical and business-based marketing and promotional materials. This book is a departure from their normal daily life but fits in-line with their passion of creating new and exciting material for others to enjoy.

Although this book likely won't win any awards or won't be on the top seller list, we hope you enjoy its story and that you find it valuable in how it teaches science and presents information in an easy-to-learn approach.

www.ingramcontent.com/pod-product-compliance
Lightning Source LLC
Chambersburg PA
CBHW060419190526
45169CB00002B/967